D1595440

Construction Zone

Backhoes

by Rebecca Pettiford

Bullfrog Books

Ideas for Parents and Teachers

Bullfrog Books let children practice reading informational text at the earliest reading levels. Repetition, familiar words, and photo labels support early readers.

Before Reading

- Discuss the cover photo. What does it tell them?
- Look at the picture glossary together. Read and discuss the words.

Read the Book

- "Walk" through the book and look at the photos. Let the child ask questions. Point out the photo labels.
- Read the book to the child, or have him or her read independently.

After Reading

- Prompt the child to think more. Ask: Backhoes are big machines. They dig and lift. Can you name other big machines that do this?

Bullfrog Books are published by Jump!
5357 Penn Avenue South
Minneapolis, MN 55419
www.jumplibrary.com

Library of Congress Cataloging-in-Publication Data

Names: Pettiford, Rebecca, author.
Title: Backhoes / by Rebecca Pettiford.
Description: Minneapolis, MN: Jump!, Inc., [2023]
Series: Construction zone | Includes index.
Audience: Ages 5–8.
Identifiers: LCCN 2021048386 (print)
LCCN 2021048387 (ebook)
ISBN 9781636908434 (hardcover)
ISBN 9781636908441 (paperback)
ISBN 9781636908458 (ebook)
Subjects: LCSH: Backhoes—Juvenile literature.
Classification: LCC TA735 .P46 2023 (print)
LCC TA735 (ebook) | DDC 621.8/65—dc23/eng/20211122
LC record available at https://lccn.loc.gov/2021048386
LC ebook record available at https://lccn.loc.gov/2021048387

Editor: Jenna Gleisner
Designer: Michelle Sonnek
Content Consultant: Ryan Bauer

Photo Credits: Vereshchagin Dmitry/Shutterstock, cover; Dmitry Kalinovsky/Shutterstock, 1; Yobro10/iStock, 3; gece33/iStock, 4; Konov/Shutterstock, 5; KVN1777/Shutterstock, 6–7, 23tr; ewg3D/iStock, 8–9, 10–11, 23br; Attapon Thana/Shutterstock, 12; dimid_86/iStock, 13; Kateryna Mashkevych/iStock, 14–15, 23bl; Vadmin Ratnikov/Shutterstock, 16, 23tl; ROMAN DZIUBALO/Shutterstock, 17; Aisyaqilumaranas/Shutterstock, 18–19; TheHighestQualityImages/Shutterstock, 20–21; maxpro/Shutterstock, 22; Yevhen H/Shutterstock, 24.

Printed in the United States of America at Corporate Graphics in North Mankato, Minnesota.

Table of Contents

Dig It Up

This is a backhoe.
It is a big machine.
It digs.

It lifts.

It moves things, too.

Workers will fix a pipe.
Mike uses a backhoe.
He sits in the cab.

cab

He drives to the job.

He lowers the stabilizers.

They keep the backhoe still.

stabilizer

He moves the arm.

It is long.

It goes out.

It goes down.

rock
bucket

The rock bucket digs.

It digs up dirt.

The big bucket lifts.
It lifts dirt.
It lifts rocks.

It puts them in a truck.

Look! It is the pipe.
The workers fix it.
Good job!

pipe

Backhoes work hard!

Parts of a Backhoe

What are the parts of a backhoe? Take a look!

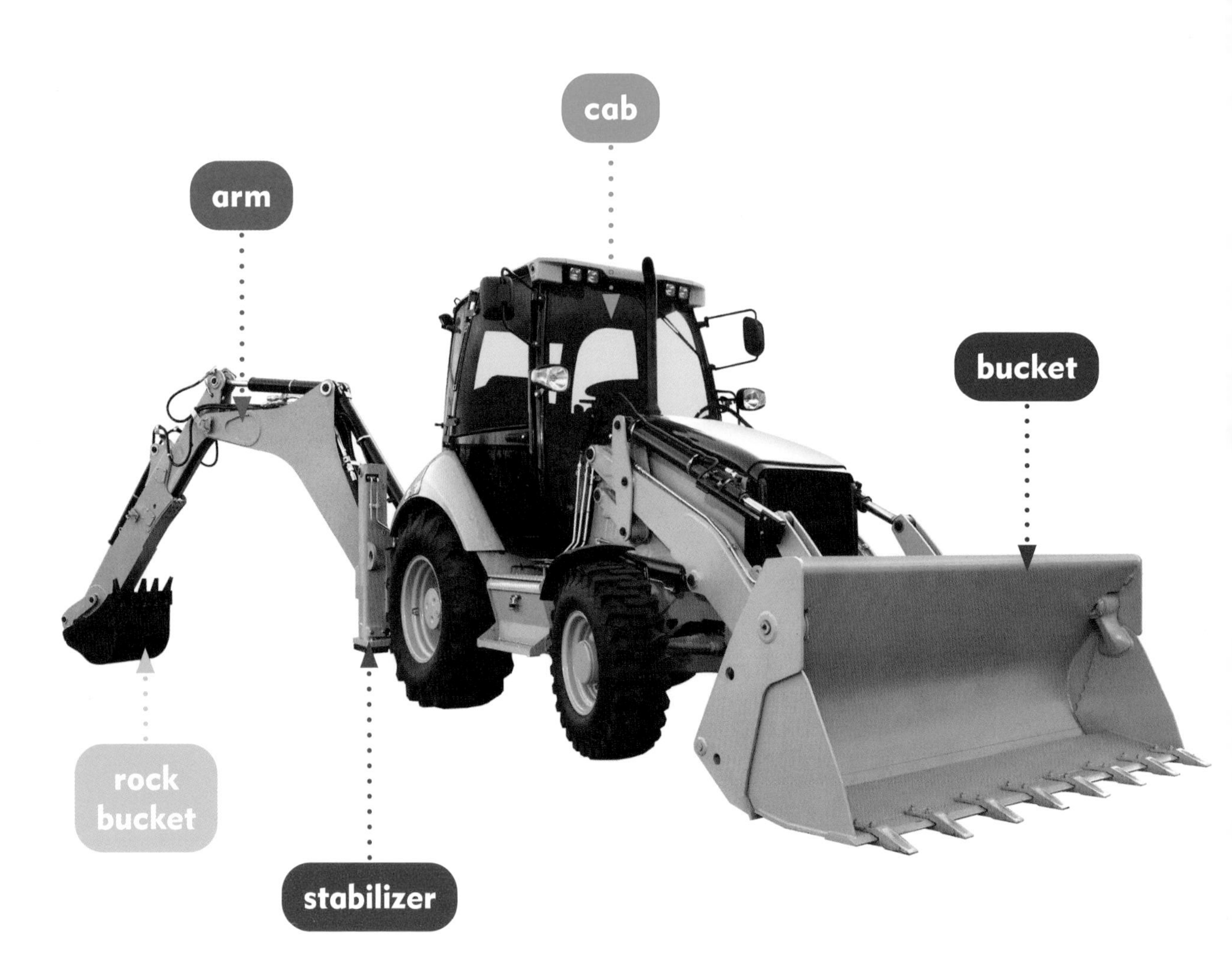

Picture Glossary

bucket
The larger bucket on a backhoe that lifts and dumps.

cab
The area in a big machine where the driver sits.

rock bucket
The smaller bucket on a backhoe that digs.

stabilizers
The two legs on a backhoe that help keep the machine steady.

Index

To Learn More

Finding more information is as easy as 1, 2, 3.

1. Go to www.factsurfer.com
2. Enter "backhoes" into the search box.
3. Choose your book to see a list of websites.